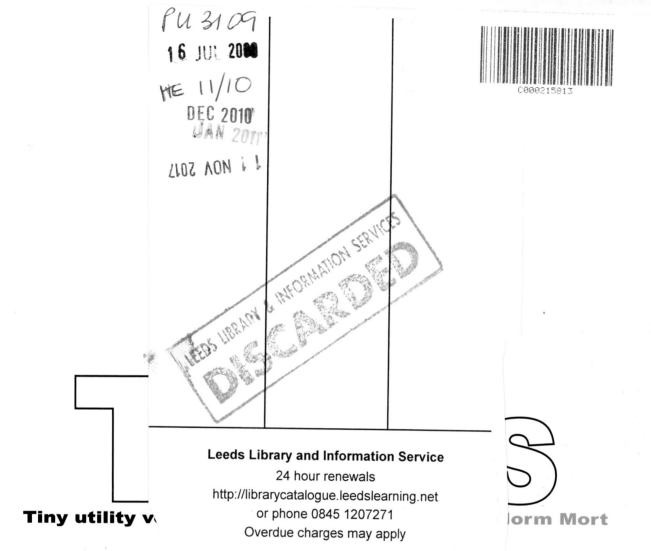

**Leeds Library and Information Service**
24 hour renewals
http://librarycatalogue.leedslearning.net
or phone 0845 1207271
Overdue charges may apply

**Tiny utility v**                              **lorm Mort**

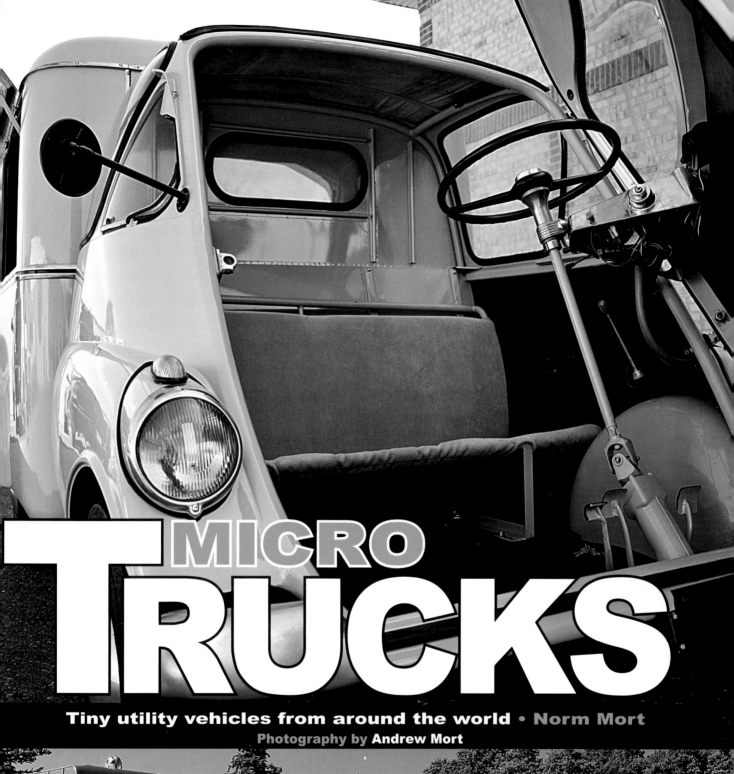

# MICRO
# TRUCKS

**Tiny utility vehicles from around the world • Norm Mort**

Photography by **Andrew Mort**

VELOCE

To my wife, Sandy, who has always encouraged this other passion in my life. *Norm Mort*

To Catherine, who completes the picture in my life. *Andrew Mort*

# CONTENTS

4

# Preface

Our son Andrew grew-up surrounded by old cars and micro cars in the garage and driveway. Our rambling old home was overflowing with books, magazines, and brochures on vintage motorcars, so it wasn't surprising that his career in photography should ultimately be linked to mine in writing.

Andrew is a recent graduate of the highly regarded Ontario College of Art and Design. Throughout his years at the college he honed his photographic skills, as well as supplementing his income. In our first book together, Micro Trucks; he combines his young, creative eye for colour and design with my nearly thirty years of writing on vintage vehicles.

Despite what we bring to Micro Trucks, this book would not have been possible without the enthusiasm,

co-operation, and vehicles of Canadian micro car and truck collector Mario Palma. Mario has imported his fascinating collection of rare micro cars and trucks from every corner of the world. As well as providing all the trucks featured, and files of information, Mario was always there to help in any way he could, from moving the vehicles to fact checking.

Thanks also go to Ford Motor Company (NA) designer and car collector Mark Conforzi, whose talent I've always admired. As a fellow Isetta owner, he was kind enough to pen the foreword.

Along the way, additional information and support was offered by micro car enthusiasts Peter Marshall and Amedeo Bellio; Fiat Topolino devotee Ezio Casagrande; librarian/archivist Laura Kotsis of the Detroit Public Library, and photographer Peter Christopher for his assistance. Much appreciation goes to John Watts of LNER who restored many of the Palma Collection vehicles, and was responsible for checking all the mechanical and technical aspects in the book. Many thanks also, to the McPhail family of Jim, Lynn and their children Riley and Maddie for their staging. Finally, much appreciation goes to Catherine Miller and Sandy Mort for their dramatizations.

**Micro trucks played an important role in the return to prosperity in the 1950s and sixties. Most performed tasks on a daily basis until repairs were no longer viable, and they were crushed. A hearty thanks to all of those who have saved these tiny icons of our heritage.**

# Foreword

1964, and although the world was in turmoil, it was a time in my young life and little neighbourhood where suddenly the past, present, and future all became one. It was a hot summer day in Toronto, and one of the fondest and probably most significant memories from my childhood. A day that not only had a monumental quality, but one which shaped the road ahead for me, as that was the day I was introduced to the BMW Isetta.

I can still recall my Grandfather driving a wet-sand-coloured BMW Isetta as I watched in amazement. Maybe it was its micro car scale, or perhaps its dream-like quality that I found inspirational, but it wasn't until my Grandfather opened the refrigerator-like front door, and I watched the steering column and instrument binnacle swing forward, that I realized how truly ingenious this machinery was. This single moment influenced my decision to become a car designer.

I'm proud to say my passion for micro cars has continued throughout my life. The BMW Isetta holds a special place and I recently restored a 1957 BMW Isetta 300.

Owning it has opened many paths of communication, not only with car enthusiasts, but with other collectors as well.

Through this hobby I've been privileged to have met Norm Mort. His enthusiasm and knowledge of micro cars and micro trucks has not only broadened my interest, but also enriched my awareness.

A book primarily dedicated to micro trucks fills a much needed void on this subject. Norm and his son, Andrew, bring to light new information and photos of the Autobianchi and Fiat vans, IsoCarro trucks, and others. These pocket-sized vehicles possess a certain innocence and embrace a toy-like quality which will evoke childhood memories for many.

**Mark Conforzi: Ford Motor Company**
**Chief Designer Vehicle Personalization**

**Mark Conforzi (right) with Mario Palma at the Palma Collection Museum.**

Dwarfed in comparison to most of the trucks that travelled our roads in the 1950s and sixties, the tiny utility vehicles still performed a yeoman's service.

# Introduction

Throughout the history of motorised transportation, engineers, designers, and mechanical wizards have constantly attempted to maximise carrying capacity, while minimizing size and cost. In most cases, these vehicles were designed to appeal to people looking for something more accommodating than a bicycle or motorcycle. Others were also seeking an increase in comfort and the ability to transport more passengers or cargo at a minimal rise in cost.

As a result, as well as micro cars there were micro trucks built for the entrepreneur or municipality that wanted to upgrade from the handcart or retire the donkey.

The amount of success in the marketplace of micro cars and trucks over the decades has always been dependent on a multitude of factors. Design was a determining issue, which included not only the mechanical configuration and engine, but also the quality of assembly, the choice of materials, and the overall reliability of the vehicle.

Even with a successful design, the manufacturer of micro vehicles has always been dependent on marketing strategies, the strength of its dealer organisation, parts availability, reliability, and major component suppliers. In some cases a successful design suffered from poor management practices, assembly techniques, labour problems or even political unrest.

However, perhaps the most significant factor has been the continual changes in economic conditions. Good ideas from five years before often floundered when a design reached a very different market. The availability of larger, more comfortable and powerful vehicles from major manufacturers at a slightly higher price impacted on sales, as did the accessibility of used vehicles and easy financing.

This is the first book focusing solely on micro trucks, but in no way covers all the micro trucks ever built. This volume covers some of the more significant micro trucks produced from 1948 to 1968. Over those two decades certain designs evolved, while others remained virtually unchanged.

Presented is a cross-section of some uncommon (Autobianchi, Moto Guzzi 500 Ercole and New Map Soylto) and some extremely rare (1954 and 1955 IsoCarro Furgone, 1961 IsoCarro 150, 1948 Fiat 500 Topolino B, 1950 Fiat 500 C Furgone) models, as well as two micro trucks that have survived in greater numbers (Piaggio Ape and Daihatsu Trimobile).

It is important to emphasise the fact that we had the luxury of working with a dedicated collector, who went to great lengths to restore his micro trucks to original factory specifications and accumulated files of information which have not been available to others. This material became the basis of my research. Yet these micro trucks were built in a less complicated time when people, not computers or robots, assembled vehicles. As a result, parts were often substituted and changes made on a regular basis to keep production moving.

Included are many photographs of the mechanical details and styling cues, complemented by some rather whimsical period-like lifestyle shots of these unique micro haulers.

**Norm Mort**
**Wellington, Ontario, Canada**

8

# 1

# ISO

In 1953, ISO opened a very different front door to transportation.

By the early 1950s ISO had established itself as a builder of motor scooters, motorcycles, and similarly-powered and based IsoCarro vehicles, as well as refrigerators. It seems that the refrigerator reference is always included in a history of ISO because of its Isetta bubble car and its characteristic single front door.

Many enthusiasts are unaware that it was ISO that designed the tiny egg-shaped Isetta and not BMW. Isetta literally translates as 'Little ISO,' but was affectionately called the 'Little Egg.'

BMW was one of many to acquire a licence to build this ingenious bubble design in its own factory. Variations of the original Isetta would also be built in Belgium, Britain, France, Spain, and Brazil.

Introduced in 1953 at the Turin Auto Show, the ISO was an immediate sensation. However, from 1953 to 1955 ISO delivered just 4900 units of its Isetta compared to the 161,728 built by BMW when it halted production in May, 1962. The argument has been put forth that it was competition from Fiat, specifically the Fiat Topolino and 600 and the makers of other small cars that ruined the company's chances of success in Italy.

More likely, it was the unreliability and unnecessary complication of the two-stroke, twin-piston ISO engine. This siamese-cylinder engine featured twin bores, with one piston acting as the outlet, and the other the inlet, all

**The delicate, jewel-like crest of the Iso Isetta was more than appropriate for such an innovative design.**

under a single head unit. Trojan trucks featured a similar engine design. Although factory prepared Isettas had fine finishes in the 1954 and 1955 Mille Miglia endurance races, everyday maintenance in the hands of owners was another thing.

The BMW Isetta was first fitted with a 245cc, single-cylinder R25 motorcycle engine of its own design. It was a four-stroke engine with slightly more power and proved to be considerably more reliable. The BMW Isetta was also extensively re-engineered to provide a more comfortable ride, improved braking, lighting and better road-holding.

The BMW Isetta was also continually upgraded and improved in every respect, including a slightly more powerful and larger engine.

At the same time, ISO didn't have the widespread dealer network and parts depots to compete with Fiat and other long-established builders in Italy.

Velam of France purchased its engine from ISO and made only a few minor engine changes, but extensively modified the overall Isetta design. Velam would go on to build 7115 French Isettas from June 1955 to January 1958, but couldn't compete in the marketplace with the Citroen 2CV or small Renault.

Conceived by ISO as a four-wheel design, it was the British who offered the Isetta with three. The British Isetta continued in production after most of the other manufacturers had abandoned the market, with the last being built in 1964.

ISO had great plans for its Isetta that went well beyond the original successful concept created by aeronautical genius, Ermenegildo Preti. Preti believed in designing small cars as small cars, and not just scaled-down large cars. It was unlikely Preti supported the idea of turning his little egg into a larger omelette. Meanwhile, the company certainly saw the possibilities of offering small ISO trucks using the Isetta as a basis.

The Isetta body was extensively modified and a

**Although the creator of the Isetta, ISO itself built relatively few of its ingenious design in either car or truck form.**

Above: More was involved in transforming the Isetta into an IsoCarro truck than just placing the standard car body on a heavier frame.

ISO built very few of its Isetta-based trucks. The most common appear to be the pickup version, but only a handful of these are known to exist.

Pictured atop this 1951 American-built Mack truck is a 1954 IsoCarro.

The IsoCarro pickups were popular in both city and rural areas.

tubular chassis added to support a platform, pickup box or cube van body.

Very limited production of these micro trucks took place in Italy.

It was commonly thought that the proposed Isetta als Kipper (dump truck), als Feuerwehr (ladder truck), AutoCarro (platform truck), and carrier van were concepts only in Italy, despite the fact that ISO printed brochures.

Some period photographs exist of an Italian Liquigas tanker, a Trateur Benito, tipper and others, but only a miniscule number of the limited number of ISO micro trucks appear to have survived.

A photograph also exists of an Italgas van, but that is thought to be a promotional photograph only and never actually built.

An ad from Autovehiculos, a distributor in Madrid, Spain, shows the full range of Isetta AutoCarro micro trucks, including a raised platform truck. Spanish production of ISO vehicles dated back to the early post-war years. The long produced, three-wheel Titan motorcycle-based truck came to a conclusion in 1958.

An estimated 4000-5000 ISO Isetta vehicles were built in Spain by what eventually became known as Borgward ISO Ecpanola SA. Yet, out of that figure, only a small percentage were trucks, just a handful of which have survived.

Back in Italy, ISO did continue to manufacture small trucks after the Isetta IsoCarro. This included the smaller 150 that was designed to compete with the Vespa Ape and Lambretta FLI, as well as a larger 400-series range. Yet ISO and owner, Renzo Rivolta, would not re-enter the car market until 1962 when the up-market GT known as the ISO Rivolta was introduced.

## 1954 IsoCarro 500 pickup

The building of ISO's micro trucks went well beyond simply attaching a rear box. Designed to carry 500kg, the basic car body underwent considerable modification. The welding marks are quite noticeable where the reshaped side windows were fitted. The curvaceous rear of the cab was cut off and replaced by a flat slab. A slightly modified version of the 236cc engine was fitted, but with the same transmission. The motor sat behind and below the cab, in the frame, just in front of the pickup box. The oil and gas mixture was added via two external filler caps on the left side.

A substantial, 68mm (2⅝in) in diameter, tubular frame was designed for the resulting stretched wheelbase of 2150mm (85in). The pickup box measured 2000mm (78¾in) in length, 1400mm (55in) in width, and 300mm (11¾in) in depth and had a fold-down tailgate.

The familiar front opening door with hinged steering

wheel provided easy step-in access to the front bench seat, like that found in the ISO Isetta.

The instrumentation was also identical and consisted of a very optimistic 100kph VDO speedometer and odometer. The sunroof and vent windows allowed for optimum circulation, as the bubbled side windows

**The ISO 236cc engine underwent only minor modifications to improve torque, and utilized the standard Isetta transmission.**

**The IsoCarro 500 pickup featured a sizable box, considering its overall dimensions.**

were fixed. A rubber mat covered the floor and basic pressboard panels were fitted on the sides and door.

Power from the tiny ISO was transferred via a driveshaft, to a solid rear axle and differential.

### Driving impressions of an IsoCarro 500

Egress and ingress into an Isetta is not as easy as it may appear and takes a bit of mastering, particularly when it comes to the smaller IsoCarro cabin. The door is not light and is held open by a pneumatic strut, but weighted and carefully angled so as to not fly open. Still, one wants to ensure it is slammed shut, chiefly because the steering is hinged to the door.

Once you've stepped inside and closed the door you appreciate the fact you can easily slide along the bench seat. Sitting behind the steering wheel of an IsoCarro 500 pickup is predictably not unlike driving an Isetta. What there is of instrumentation can basically be ignored as breaking the speed limit won't be a concern. The remote

**Egress and ingress are not as easy as they may appear.**

Added length did nothing to alleviate that 'tippy' feeling when driving an IsoCarro at speed – even if that speed was just 35mph!

**Above right: Stylish front fender running light enhances the inherent charm of the IsoCarro 500 pickup.**

**Above left: Although the body was modified extensively on the IsoCarro, the familiar refrigerator-style door was retained.**

## Specifications
Manufacturer: ISO SpA, Milan, Italy
Model: 1954 IsoCarro 500 Pickup
Weight: 475kg (1047lb)
Max. weight: 1125 kg (2480lb)
Load capacity: 500kg (1102lb)
Engine: 9.5hp, 236cc, single cylinder, air-cooled, 2-stroke, twin-piston
Transmission: Four-speed & reverse
Starter: Electric
Steering: Worm
Brakes: Hydraulic
Electrics: 12 volt
Body: Steel
Chassis: Tubular
Suspension: Front: Dubonnet-type, independent
　　　　　　Rear: semi-elliptic with shocks
Wheels: 500x10
Top speed: 85km/h (53mph)

shifter on your left side, protruding through a hole in the panel has a shift knob smaller than the knob for opening the door. On the dash to the right is the diminutive wiper motor for working the single wiper.

The steering column is fitted with a unit to work the lights and signals.

The sunroof is held in place by two sliding bolts with Bakelite knobs. When released the top folds back on itself in a framework. The opening is split into two by a metal bar, placed there more for the additional rigidity of the cab than to keep the top from sagging.

The cab is like a roll cage. Framed in unpadded tubes, the body panels are welded into place. The Plexiglas windows are framed in aluminum. The side windows are fixed and bubble out. The large front windscreen is glass.

The IsoCarro's shortcomings become immediately apparent when driving. Apart from the noise factor, there is very little power.

Mario Palma has found that no matter how much you adjust the fuel mixture clouds of smoke billow out. Other observations include, "It's a bit scary going downhill and when cornering; despite its size, the IsoCarro has that 'tippy' three-wheeler feeling. I also make it a practice to always take a passenger to push. Still, it's a thrilling ride and unique.

## 1955 ISO Furgone van

The existence of any Italian-built ISO Furgone was doubtful, so finding two was remarkable.

Shipped from Italy in 2001, one was very complete and fully restored to exact specs. Other than the cube body it was identical to the IsoCarro pickup truck. Access to the cargo area was through a large right-hand hinged

**The Italian IsoCarro Van was believed by many to be only a concept and never built.**

Access to the van cargo area was via a large
rear door.

As well as a rear door the van featured a
hatch window on the side.

**This IsoCarro began life as a van, but somewhere along the line the cargo body was cut up. A replica cube body was built, but, as a replica, was converted to a period emergency vehicle. It is powered by an ATV engine for more practical everyday driving.**

rear door. There was also an opening on each side; one with a Plexiglas window fitted, and the other, a hatch panel.

The other 1955 IsoCarro van had no rear cube fitted at time of purchase, but was verified as a Furgone version. The cube bodywork was available in a variety of styles. This example was completely rebuilt to look as close to an original as possible, but as a period emergency vehicle. To make it a more practical, everyday driver this Furgone was fitted with a modern four-stroke ATV engine and better brakes. The emergency light on top was created in a fanciful fifties style for a period look.

### IsoCarro150/C Pickup

The IsoCarro150 Pickup pictured was one of an estimated ten exported by ISO to the United States in 1961 to test the American market.

Ultimately, it was sold to an ESSO gasoline station owner located in Stroudsburg, Pennsylvania who felt the little truck was a perfect promotional vehicle. At that time ESSO was promoting its fuel with the slogan "Put a tiger in your tank!" The IsoCarro was painted in an attention-getting tiger motif. For the next twenty-seven years the 150 performed basic duties and acted as an advertising attraction at the ESSO service centre.

With a bed size 1400mm (55in) in length, 1310mm (51½ in) in width the 150/C was more than adequate to carry gas cans, oil products, and other service centre items to motorists who had broken down nearby.

When the business was sold in 1988, the owner stored the IsoCarro in his garage where it sat unused for the next eleven years.

Having been driven less than 1000 miles, it was finally sold to another Pennsylvanian enthusiast before becoming part of the Palma collection.

The two doors on the cabin design were suicide-style, with fixed side glass as only the rear-hinged vents open. The dash was very similar to that of the Ape model, with a single Vereinigte Deuta-Ota (VDO) combination speedometer and odometer, but with shifting marks. Recommended shifts for the four-speed gearbox were at an indicated 7mph, 16mph and 25 with a suggested top speed of 37mph in fourth. The wiper motor was mounted at the base of the screen, in the centre of the dash, to operate the single blade.

There were no interior panels and the seat was a simple bench with no seat back fitted or needed, since leaning forward provided greatly improved visibility and avoiding the necessity of wearing a protective helmet.

The steering was off-set which allowed for a passenger, which was convenient in a service vehicle.

The 150/C was started from outside. The gravity

**Perhaps "Put a Tiger in your tank!" was a bit much for a tiny ISO 150/C.**

**Specifications**
Manufacturer: ISO SpA, Milan, Italy
Model: 1961 IsoCarro 150/C Pickup
Length: 2770mm (109in)
Wheelbase: 1816mm (71½in)
Width: 1473mm (58in)
Height: 1840mm (64½in)
Track: 1090mm (43in)
Weight: 495kg (1092lb)
Load capacity: 350kg (772lb)
Engine: 6.5hp, 148.7cc, single cylinder, air-
    cooled, 2-stroke
Transmission: Four-speed & reverse
Starter: Crank
Steering: Handlebar staggered
Brakes: Hydraulic
Electrics: 6 volt
Body: Steel
Chassis: Tubular and channel frame
Suspension: Front: shock/coil spring
               Rear: Semi-elliptic spring
Wheels: 4.80x10
Top speed: 60km/h (37mph)

**The rear brakes are hydraulic and activated
by a foot pedal.**

Above: Fixed side glass would not have been suitable for hot North American summers.

Below: The shift points on the speedometer were unnecessary as the high engine revs soon conveyed the same message.

Although tiny; particularly in comparison with the full-size American pickup trucks, the 150/C was still more than adequate as a service vehicle.

**Torpedo-style running lights and turn signals add a dash of style.**

fed gas tank had to be opened before the left door was opened and a handle on the floor, alongside the seat was pulled back to fire the two-stroke, single cylinder engine.

Shifting was via a toe-heel lever. There was a reverse lever which allowed for four reverse speeds.

The rear brakes were hydraulic and activated by a foot pedal, while the front mechanical brake was controlled by a lever on the right handlebar. An emergency brake was to the left of the driver's seat.

The spare tire was located at the rear under the box of this very original 150/C.

Most likely there are more ISO 150 models that exist, but this is the only known example to survive other than one example of the version that wasn't fitted with doors.

## Micro car styling with truck utility

Post-war Japan's economy suffered from a recession, high inflation and meager wages. The fledgling Japanese automakers of the 1920s and thirties had never come close to the production volumes reached in North America and Europe. For example, in 1929, of the 34,843 vehicles sold in Japan, 29,388 were assembled, 5,018 were imported, and just 437 were built by Japanese industry. Also, the number of passenger cars available decreased throughout the 1930s due to government involvement.

The very first all-Japanese auto show didn't take place until 1954. Yet, at that time, the three most important status symbols in Japanese society were the electric refrigerator, washer, and vacuum cleaner. The average automobile cost as much as a house.

Although 267 motor vehicles were on display at the 1954 exposition, only 17 were passenger cars. Most were commercial vehicles, motorized bicycles, and, motorcycles.

Micro or 'midget' vehicles in Japan were known as keijidosha, kei cars, or K-Class vehicles.

These vehicles built following WWII were powered by engines of less than 360cc, and were the most popular form of transport for three reasons: poor economic conditions, stiff regulations governing mandatory regular overhauls, and the compulsory designated parking based on where you lived.

**Despite the frowning look of the Daihatsu Trimobile, it quickly brings smiles to the faces of those who see it today.**

In 1958 Daihatsu introduced its newest line of small truck known as the Trimobile.

The roots of the Daihatsu company as we know it today were originally founded by a group of individuals back in 1907.

The Hatsudoki Seizo Co Ltd of Osaka, Japan was established as a manufacturer of internal combustion engines for the domestic market. Vehicle production didn't commence until 1930. The company's first production vehicle was the 500cc engine, Model HA motorcycle-based pickup with two rear wheels.

Like many vehicle manufacturers, Daihatsu produced a wide variety of motorcycle-based transporters during the 1930s. Daihatsu often utilised proprietary engines with some of the more powerful three-wheel vehicles during this period being fitted with BMW motorcycle engines.

Production of three-wheel vehicles became an integral part of the company. At the same time there were four-wheel trucks also being developed. In April 1937, a small four-wheel truck called the Model FA was announced.

As the decade closed production began to centre on the building of 4x4 military vehicles.

Following WWII, the production of three-wheel vehicles resumed. Some of the larger versions were powered by a 1005cc engine and capable of carrying payloads as high as 750kg.

The company name was changed to Daihatsu Motor Co Ltd in 1951.

The name 'Daihatsu' stemmed from the combination of the first Japanese character used in the spelling of Osaka and the first symbol of the word 'engine manufacture.' Combined, the pronunciation in Japanese was 'dai hatsu.'

The 1951 change in name coincided with the launching of the company's first car. While the two-door Daihatsu Bee sold in relatively small numbers, the production of three-wheel utility vehicles continued to expand.

A small, three-wheel, Piaggio Ape-looking 'Midget' joined the line-up in 1957. Sales of the Midget, single-person vehicle with its 249cc engine were brisk thanks to its small turning circle and easy manoeuvrability.

In 1958, a new line of small utility models known as the 'Trimobile' or sometimes advertised as 'Tri-Mobile' was introduced.

Mechanically and in design, the Daihatsu Trimobile had evolved from the company's earlier motorcycle-based vehicles, but its front end styling bore a closer resemblance to the Bee passenger car, as did its driveline.

**The Trimobile name was certainly not the most creative aspect of this Japanese design.**

**The overall dimensions of the tiny Trimobile are quite apparent from this perspective.**

The Trimobile – so called due to its single front wheel and two rear wheels, was soon offered in a full range of body styles.

In 1959, this included the 'Bulldog' tractor for hauling multiple small trailers of light goods; the 'Exurbian' utility wagon with canvas roll-up sides and a metal roof; the 'Wagon' type with a full canvas roof; the 'Triton' metal roof pickup; the 'Courier' cube-style enclosed van; the 'Handiwagon' canvas top with an all-metal cab; and the 'Pivoteer' one-piece body, full panel truck.

The Trimobile was also being offered with a choice of two, single-cylinder, two-stroke engines.

The base engine was a 10hp, 249cc air-cooled powerplant or you could opt for the slightly more powerful 12hp, 305cc engine.

In 1959, both the Trimobile and the Midget were being exported, and gained popularity in the United States, Indonesia, and other countries.

**The side windows simply clipped into place when required.**

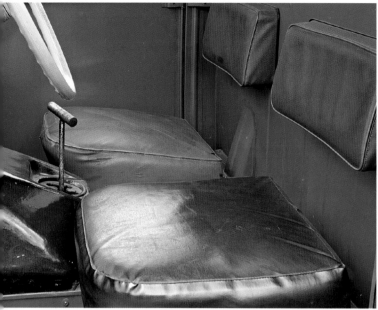

**The two removable front seats were separated by the handbrake.**

By 1965, Daihatsu was building everything from tiny three-wheel vehicles and sub-compacts to half-ton and two-ton trucks, as well as buses, and construction equipment. In addition to vehicle production, the company manufactured giant diesel generators in Japan and could boast of a presence in eighty markets throughout the world. Production of the Trimobile ceased in 1975.

### 1959 Daihatsu Trimobile

This very original 1959 Daihatsu Trimobile pickup was purportedly used as a fish truck on the US west coast for years.

It featured a vinyl, snap-on sunroof and suicide doors with permanently affixed at the bottom, vinyl and plastic side curtains that hung down on the inside of the panel-less doors when not in use.

The windscreen and two rear windows in the metal cab were made of Plexiglas.

The lift-off, two wooden-framed seats were separated by the handbrake. Additional storage for tools, etc was housed under the driver's seat, while under passenger seat sat the battery and starter motor. Removing the seat cushions also allowed for easier access to the engine which could be accessed from the interior. The seat squabs consisted of two separate, rectangular, curved pads bolted to the rear of the cab wall.

The dash contained no instruments, but simply an ignition key, starter button, toggle switch for the two headlamps and running lights, and an oil pressure

**With little inside cab storage area, the permanently affixed at the bottom vinyl and plastic side windows just hang down on the inside of the doors.**

warning light. In front of the driver was a two-spoke steering wheel with a single grab handle affixed to the dash in front of the passenger.

Although small, the interior was not cramped and provided better than average comfort for a micro truck.

A combination engine and transmission metal shroud separated the two footwells and was easily lifted off with the removal of four bolts and three screws. The gearshift handle also had to be unscrewed before the engine cover could be removed. Once the shroud was eliminated, the servicing of the motor was fairly straightforward.

The single cylinder engine was bolted into a frame, just in front of the knees. The entire placement was not unlike that of the Reliant Regal and others. Running down the centre of the narrow ladder frame was the driveshaft leading to a differential. The rear box in this pickup truck version measured 1016mm (40in) in length, 1066mm (45¾in) wide with a drop-down tailgate.

**Specifications**
**Manufacturer: Daihatsu Kogyo Kabushiki Kaisha**
**Model:1959 Daihatsu Trimobile**
**Length: 2885mm (114in)**
**Wheelbase: 1810mm (71in)**
**Width: 1296mm (51in)**
**Height: 1510mm (59in)**
**Track: 1122mm (44in)**
**Weight: 412kg (908)**
**Load capacity: 350kg (772lb)**
**Engine: 12hp, 305cc single-cylinder, two-stroke**
**Transmission: 3-speed & reverse**
**Starter/drive: Electric**
**Steering: Direct**
**Brakes: Rear rod brakes**
**Electrics: 12 volt**
**Body: Steel**
**Frame: Ladder**
**Suspension: Front: dual coil springs and shocks; Rear: semi-elliptic springs**
**Wheels: 5.00 x 9**
**Top speed: 65km/h (40mph)**

**The pickup box was designed for small loads only.**

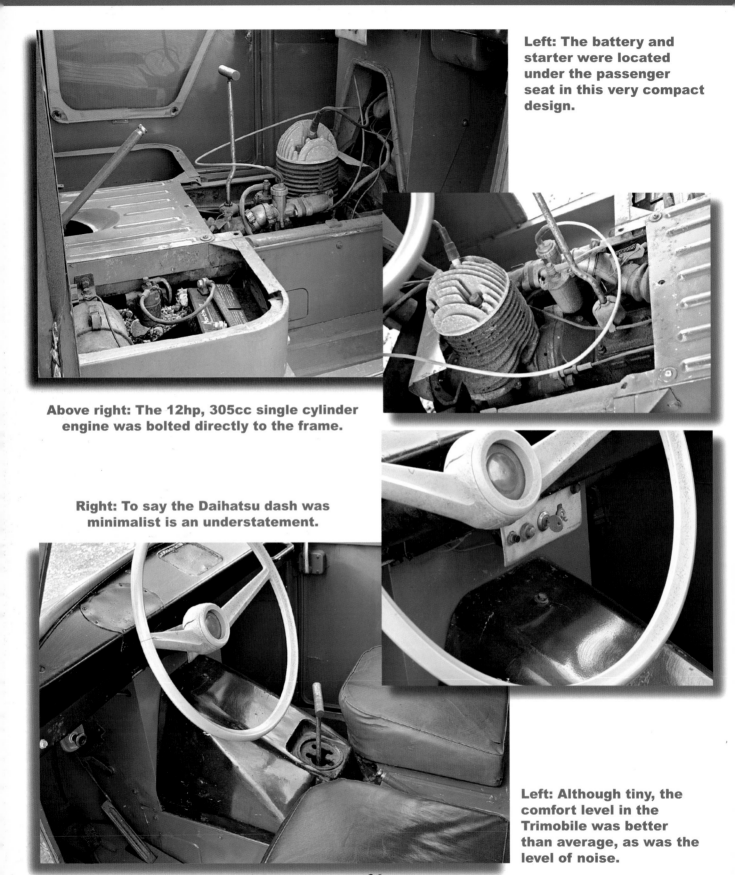

Left: The battery and starter were located under the passenger seat in this very compact design.

Above right: The 12hp, 305cc single cylinder engine was bolted directly to the frame.

Right: To say the Daihatsu dash was minimalist is an understatement.

Left: Although tiny, the comfort level in the Trimobile was better than average, as was the level of noise.

36

With its tiny vent on the front hood, dropdown side Plexiglas windows, and large sunroof, the driver and passenger in a Trimobile would have felt the wind in their hair.

# 3 Moto Guzzi Ercole

**A powerful Ercole for hauling big loads**

Moto Guzzi was founded by an Italian Air Force pilot, Giorgio Parodi, along with mechanic and driver, Carlo Guzzi, at Mandello del Lario. The company's eagle logo was taken from the Italian air force in honour of their friend and associate, Giovanni Ravelli who was killed during WWI.

As the 1920s evolved, Moto Guzzi soon became known for its design and quality construction.

In fact, over the years the Moto Guzzi name would

The Moto Guzzi 1500kg Ercole remained in production until 1980.

**The cab of the Ercole is basically designed to protect the driver from the harshest elements only.**

**Despite its overall diminutive size, a Moto Guzzi Ercole was a substantial vehicle.**

become synonymous worldwide with fast, precision built motorcycles with a legendary racing record. Yet the firm also had a long standing history of building three-wheel utility vehicles. Dating back to 1928, when it introduced its Tipo 107 Motocarro, these micro trucks played an integral role in the financial well-being of the company at various times in its history, providing excellent fuel economy. These small utility vehicles fulfilled essential tasks and were often the choice of Italian merchants, tradesmen, public works utilities and the military.

When introduced, the Tipo 107 Motocarro was powered by its already famous single-cylinder 498.4cc engine that dated back to the company's first venture into the motorcycle business. Unveiled in 1920, the Guzzi e Parodi caused an immediate sensation. The engine was of such advanced design it would continue in production until 1976. What was even more remarkable was the fact that when production ended, it retained the same bore and stroke dimensions of the original.

The Moto Guzzi Motocarro was a motorcycle-type, three-wheel delivery with power to the rear wheels through its horizontally mounted engine. This would be

**Above: The two suicide doors are locked into place with the most primitive of latching systems, yet the overall design of the Moto Guzzi Ercole is filled with examples of complexity and creative engineering.**

**Left: Everyday driving of a Moto Guzzi Ercole would have been a challenge for most drivers used to the comforts of today's trucks.**

a characteristic of all the company's larger three-wheel vehicles.

In 1938, Moto Guzzi revealed its 499cc (30cu in) Model ER Motocarro, consisting of a combination of the front half of a tubular motorcycle frame, with truck-inspired box frame at the rear.

From 1938 onwards, the shaft-driven Motocarro, equipped with a four-speed transmission, was capable of hauling 907kg (2000lb). During the 1940s the Trialce was built solely for use by the Italian army.

From 1938, until production was halted in 1942, there were 5143 ER Motocarro versions built.

Following WWII, the 1500kg Ercole was introduced and later equipped with a 5-speed transmission with reverse. The large, tough Ercole versions ceased production in 1980.

**The side Plexiglas windows simply slid down for fresh air.**

From 1956 to 1970, Moto Guzzi also sold an Ercolino Motocarro powered by its 192cc, single-cylinder engine from the Galletto (Cockerel) motor scooter. Available in open or cab versions, it had a load capacity of 350kg (772lb) and was ideal for the narrow streets in Italian towns and cities. It was Moto Guzzi's attempt to compete with Piaggio, Lambretta, and ISO in the Italian market. Later models were capable of hauling up to 590kg (1300lb).

Also, in 1962, Moto Guzzi introduced the Aiace Motocarro with a fully enclosed cab and powered by the Zigolo 110 engine. The three-wheel, scooter-like Dingotre was another open goods carrier, built from 1965 to 1968. Smaller still were versions of the Moto Guzzi Motozappa range of farm vehicles, roto-tillers and cultivators. As well as being capable, lightweight cultivators, a small box was fitted for hauling.

Moto Guzzi also continued with its military connection, and between 1959 and 1963 delivered 220 examples of its 3x3 vehicle powered by a 750cc transverse V-twin engine. It garnered great publicity when the cart proved it could climb a brick wall.

### 1959 Moto Guzzi Ercole Tilt-Bed

The Moto Guzzi Ercole design consists of a substantial rear frame, welded to the reinforced front-half of a motorcycle frame. The front shell or cab, which is basically only a protective skin from the elements, is braced at the rear and bolted through the floor to frame extensions. The shell of the cab is held together with metal strips.

This spacious cab is fitted with two suicide doors and simple aluminum door handles working a rather primitive latch system. There is no interior padding or sound-proofing in the cab. The only thoughtful touch, from an operator's point of view, is a metal pocket on the back wall designed to hold work orders. A single wiper for the flat glass windscreen, along with one front headlamp was standard. There was no front bumper, but then there was no sub-frame where it could be secured.

All motorcycle inside – there was plenty of elbow room, once positioned on the customary saddle seat and gripping the handlebars. The floor could withstand the weight of the driver, but did not extend all the way over to the frame of the motorcycle. This allowed for easy maintenance, but the downside was rather drafty, somewhat exposed driving. Wet weather would be particularly unpleasant, as would the fumes that tended to build-up inside.

For additional fresh air the side Plexiglas windows slide down into door frames. The vent windows were permanently fixed.

Cycle fenders were fitted over all three wheels. The front wheel was standard motorcycle, but the larger rear wheels were definitely truck. The Moto Guzzi Ercole 500 was designed for hauling heavier loads than the majority of micro-sized Italian trucks. This example featured a manually-operated tilt bed measuring 2184mm (86in) long, 2083mm (82in) wide and 394mm (15½in) deep.

The single spare was housed under the bed at the rear. (This makes one wonder how front tyre failures were handled.) A five-speed transmission transferred power to a driveshaft and rear differential from the 498cc (30.3cu in) air-cooled engine. The standard motorcycle fuel tank was supplemented by an auxiliary tank.

Built from 1946 to 1980, the total production of all Moto Guzzi Ercole Motocarro 499cc (30.5cu in) versions is estimated at over 38,000 units. Although this model type was built from

1950 to 1959, most were open versions with no cab. It is believed that some Ercole chassis were sold without any bodywork to the military and others fitted with special bodywork by outside firms.

**The front wheel of the Ercole was standard motorcycle.**

**Far left:** An hydraulic system was available, but this example was fitted with a manually-operated tilt.

**Above:** A gated, five-speed transmission was standard on later models.

**Left:** A truck-like rear end made this 3-wheel micro a big hauler.

## Specifications

**Manufacturer:** Moto Guzzi SpA, Mandala del Lair, Como (1928-1967)
**Model:** Moto Guzzi Ercole 500 Tilt-Bed Pickup
**Length:** 3505mm (138in)
**Width:** 2083mm (82in)
**Height:** 1842mm (72-1/in)
**Wheelbase:** 2440mm (96in)
**Track:** 1270mm (50in)
**Load capacity:** 1500kg (3307lb)
**Engine:** 17.8hp, 499cc (30.5cu in), Falcone two-stroke

**Transmission:** 5-speed
**Starter/drive:** Electric
**Steering:** Direct
**Brakes:** Hydraulic rear
**Electrics:** 6 volt
**Body:** Steel
**Chassis:** Steel frame
**Suspension:** Front: coil
                         Rear: Semi-elliptic springs
**Wheels:** Front: 120/90-17 Rear: 400-19
**Top speed:** 60km/h (40mph)

**There was no doubt that the basis of the Moto Guzzi Ercole's design stemmed from a motorcycle.**

## France's long-standing, very basic micro workhorse

Despite the relative obscurity of its name today, New Map was a long standing Lyons, France-based company known for its motorcycles, scooters, tiny cars, and utility trucks. It was a diverse operation from the start and was destined to become even more so.

Founded in 1920 by Paul Martin, over the ensuing decades, New Map concentrated mainly on the building of motorcycles and scooters. These two-wheelers were powered by proprietary engines ranging in size from 98-998cc (6.61cu in).

In 1938, the company introduced its first car under the Rolux name manufactured by a sister company. This stylish, two-seater, pre-war roadster was one of the first in a new wave of micro cars that started to appear in the depression-ridden thirties. This first Rolux model was powered by a 100cc, single cylinder, air-cooled, two-stroke Fichtel and Sachs engine placed in the rear. Rolux built an open version and, reportedly, a commercial vehicle.

A 1940 re-organization of the company had seen Robert Robin effectively replace Paul Martin as head of New Map's production facilities.

Following WWII, New Map focused on the under- 250cc scooter market, but also re-introduced an up-dated Rolux to compete in the latest micro car boom.

In 1947, the New Map division, now known as Societe Rolux in Clermont-Ferrand, built the Rolux Baby VB 60, which featured a single-cylinder 5hp, 125cc (7.6cu in), Ydral air-cooled engine mounted in the rear.

**The Solyto came in a full line of body styles to fit the needs of both city merchant and farmer.**

In 1949, a more powerful Rolux Baby VB 61 was announced fitted with a 6hp, 175cc (10.6cu in) two-stroke, air-cooled engine.

By 1952, about three hundred Societe Rolux vehicles had been built and overall an estimated, grand total of about one thousand Rolux cars rolled off the New Map assembly line from 1938 to 1952.

New Map then halted Rolux production in favour of its more popular scooters and a range of three-wheeled goods-carrying vehicles.

The New Map three-wheel Solyto micro trucks were built by the company's sheet metal division, known as the Societe Lyonnaise de Tolerie. The unusual Solyto nomenclature had originated through the use of the first two letters of each word in the company name.

The first Solyto triporteurs appeared in 1956. Unlike many of the other scooter-based and powered utility vehicles of this era from other manufacturers such as Vespa, Lambretta, Mazda, Daihatsu, etc., the New Map Solyto was fitted with a steering wheel rather than the traditional handlebars to turn the single front wheel.

A separate frame allowed for the fitting of different bodies to suit the needs of both farmers and business people.

The three body styles offered featured a metal framework clothed in steel panels, wood, and material. The rear floor was made of wood and the top was a combination of metal and cloth. You could order your Solyto fully open or partially enclosed, depending on the body style chosen.

The Fourgonette Tolee (van) and the Break-Camping (window van) sported cube-like enclosures at the rear, while the Fourgonette Bachee was fitted with a full vinyl/canvas roof. Interestingly, the soft-top versions featured suicide doors, whereas the Break-Camping hardtops were hinged at the front.

In both cases, the cargo area was a generous 1050mm (41in) long, 1030mm (40in)

**A steering wheel instead of handle bars was an appreciated design feature.**

wide and 1100mm (43in) high, which was substantial considering the Solyto's overall dimensions.

A detachable drop-down tailgate, just 420mm (16in) above the ground made the Solyto micro trucks easy to load.

Originally the Solyto was fitted with a two-stroke, 125cc Ydral, air-cooled engine mounted on the front wheel (1956-1959). This power plant later gave way to a similar sized Ultima engine, offered with an automatic transmission and in its final years a KV unit.

As the 1950s progressed, the highly competitive motorcycle market in Europe faced new challenges from Japanese manufacturers. Honda, Suzuki, etc. began exporting in large numbers to meet the subsequent increase in demand.

At the same time, the motorcycle and scooter market was shrinking as more and more new, economical mini cars were being offered by the large volume manufacturers and less expensive, reliable used cars became available.

In Europe, the Piaggio Vespa had become an overnight sensation, and by 1959 New Map was forced to close its motorcycle and scooter division. As a result, the Solyto micro trucks were the only vehicles being offered by this once booming concern. The company's ultimate survival was achieved through further manufacturing diversification over the next decade.

Throughout that period, New Map had supplied a variety of metal fabrications to the Societe Anonyme de KV, in Chassieu. KV had principally built telephone equipment, but in 1970 offered its first micro car.

In 1970, KV took over New Map and continued with the production of Solyto micro trucks. By 1971, the Solyto was being sold as the TC-8 by KV.

Joseph Spalek, who had designed the automatic

**Above: Storage space was generous, regardless of the model you chose. The enclosed vans provided greater protection for more expensive and delicate cargo, whereas the open versions offered more load flexibility.**

transmission used in the Ultima-powered Solyto in 1965 had become the firm's director and continued in that position until Solyto production ceased in 1974.

In total, it is estimated that only 4000 Solyto micro trucks were built over the 18 years of production.

In 1978, KV was sold and became KVS under the guidance of former New Map executive, Joseph Spalek – hence the added 'S'. Spalek focused on the production of a range of KV micro cars.

As for KV, about 2000 of these all-metal, angular micro cars were built from 1970, until the firm finally closed its doors in 1989.

### 1956 Solyto Break-Camping

This fully restored example was known as a Break-Camping version, or what would commonly be termed a window van.

The instrumentation, on the near horizontal dash, consisted of only a speedometer with an odometer

**The first Solyto models were powered by a two-stroke, 125cc Ydral engine.**

reading. The amount of gas left in the tank was easily checked as the gas tank and cap were built into the dash, near the door on the passenger side. Starting was via a rope pull; the handle dangling from a hole on the top of the dash.

The gearshift handle stuck out from the centre of the dash, and was not so conveniently placed between a metal access cover and the steering wheel and steering column.

A very bus-like steering wheel was fitted rather than handle bars, despite the fact the front wheel sat between forks. The steering was lubricated with oil that was filled via a shaft with a cap on the centre of the dash.

The 1956, New Map Solyto was fitted with bucket seats of a very primitive design. These consisted of a tubular frame which supported a hammock-like seat made of canvas.

Access to the rear cargo area was via two doors with windows that opened 180 degrees. An entirely flat floor allowed for transporting large, albeit fairly light objects such as rattan furniture.

Access to the engine of this front engine, front-wheel-drive micro truck, was via a front grille panel that was held in place by clips and two knobs that unscrew. With the panel lifted off and put to the side, the 5.5hp, 125cc Ydral single-cylinder, two-stroke engine sitting between the two forks was easily serviced.

The Solyto was popular, both as a rural and city vehicle because of its light weight, manoeuvrability and tiny dimensions. At the same time it provided a

**Designed more for hauling loads than for comfortable transport, the cab area was made as small as 'humanly' possible.**

**Right:** The two rear doors with windows opened 180-degrees for easy loading onto an entirely flat floor.

**Left:** The 5.5hp, 125cc Ydral single-cylinder, two-stroke engine sitting between the two forks was easily serviced.

**Below:** Access to the engine required removal of a front panel.

Solyto vans such as this one rolled along
on tiny Michelin 270 x 90 tyres.

Above: The vinyl top that was held in place by a series of elasticized cords which fastened around wooden spools bolted onto the sides of the metal body.

Left: Tubular framework added to the overall rigidity of the New Map Solyto, but there was never any mention as to any safety aspect in the event of a rollover.

Above: These tail lights are believed to be original. The simple, yet effective design made replacing bulbs an easy task.

Left: A low floor level and drop down or detachable tailgate made for easy loading.

Below: There was no fancy badge on the back of a Solyto, just a simple decal bearing its name.

comparatively large cargo area and excellent fuel consumption.

Capable of a top speed of 50km/h (30mph) the Solyto consumed merely 4-litres of fuel per 100km (62 miles) for a resulting range of 300km (185 miles) between fuel stops.

Solyto advertised the fact that the wide rear track on its trucks provided the necessary stability for transporting loads safely and comfortably. Those traits, plus the fact it was simple and economical to operate and could carry substantial loads, were a key reason for its niche market survival for over two decades.

## The 1958 Solyto Fourgonette Bachee
This 1958 Solyto, with its full vinyl/canvas roof was known as a Fourgonette Bachee. It featured a fixed tubular framework, which supported a vinyl top that was held in place by a series of elasticated cords, fastened around wooded spools, bolted onto the sides of the metal body. The tubular framework was strengthened on the side by lengths of V-shaped pieces of metal. These supports – two on each side and four across – provided the necessary structural integrity. The front of the canopy was fastened to the windscreen by small springs. There were two holes on the top of the A-posts for adjustment. Fastened in this way, the top on the Solyto would not rip off at maximum speed or traveling downhill.

A French claxon horn and Marshal headlamps were fitted.

On this version, a quarter-elliptic spring was attached to the right side of the wheel to cushion the front end. (Some versions I've seen of the same year were fitted with a shock absorber and coil spring.)

This 1958 example also featured tail light lenses secured by metal bands which clipped into place. This made it easy to change bulbs, while offering additional protection to the tail light from drivers who park by bump.

**Specifications**
Manufacturer: New Map
Model: 1956 Solyto Break Camping
Length: 2450mm (8ft)
Wheelbase: (58in)
Width: 1280mm (4ft 2in)
Height: 1450mm (4ft 9in)
Track: 1120mm (3ft 8in)
Weight: 200kg (441lb)
Storage capacity: 1189cc (73cu in)
Engine: 5.5hp, 125cc Ydral single-cylinder, two-stroke
Transmission: 3-speed
Starter: Rope pull
Steering: Direct
Brakes: Cable
Electrics: 6 volt
Body: Steel/wood
Construction: Rail frame
Suspension: Front: Shock/rubber bushing
                       Rear: Leaf spring
Wheels: 270 X 90
Top speed: Advertised 50km/h (30mph)

The flexibility of a canopy also allowed the enjoyment of good weather on hot, sunny days.

# 5 Piaggio Ape

## The Ape that conquered the World!

Enrico Piaggio purchased land with the idea of building a company to supply the nearby shipyards with finished lumber in Genoa, and was later joined by his son Rinaldo.

In 1887, Rinaldo left his father's company to establish Piaggio and Company in Genoa, to supply the needs of the naval industry. By 1889, success had led to expansion into the rail sector.

The Piaggio firm continued to evolve and entered the aircraft business in 1916. The company continued to flourish throughout the 1920s, and, by the late 1930s, was supplying products to every area of the transport industry, including trailers, tramways, trucks, and trolleys throughout Italy and other parts of Europe.

Following the death of Rinaldo Piaggio in 1938, the company became the responsibility of his two sons, Armando and Enrico.

Under their direction, Piaggio became involved in the design and construction of the first modern helicopter.

Today, the Piaggio Group, based in Pontedera (Pisa), Italy, is a global transport company and world leader in the building of two-wheelers and small, commercial 3-wheel vehicles such as the Ape.

Its worldwide manufacturing centres are also involved in building everything from planes to cars.

The Piaggio micro trucks, including the popular Ape, are also built throughout the world in such places as India (Piaggio Vehicles Private Ltd or PVPL), and in China (Piaggio Zongshen Foshan Motorcycle).

When launched in 1947, the Vespa name quickly became synonymous with motor scooters in Italy. 'Vespa'

Piaggio; like most Italian marques, has always taken pride in its badges and nameplates. This highly stylized script could easily be mistaken for something found on a more expensive make.

in Italian means 'wasp,' which aptly describes the shape of the Vespa motor scooter with its rounded front and elongated, bulging rear.

Like many of the motor scooter and motorcycle manufacturers after WWII, Piaggio also saw the need for a three-wheel commercial vehicle to transport goods and wares short distances at a minimal cost.

Being small for easy manoeuvring through crowded streets, these micro transporters required only minimal power. Scooter engines were ideal and Piaggio, which

had leapt to the forefront of that market immediately, applied its knowledge to designing a three-wheel commercial vehicle.

It was obvious that the resulting design was ideally suited as a delivery, pickup, sanitation, construction, commercial or public utility vehicle.

The company's first version appeared in 1946. It was a traditional design with a box placed in front over two wheels, and a seat just forward of the single rear wheel.

The 49cc motorcycle-powered Tri-Ciao was definitely more scooter in its design than truck. Loads were limited in size just from the viewpoint of visibility. This similar, but smaller, commercial vehicle was introduced in 1947, and would soon lead to numerous other model lines, including the popular Ape.

The more practical and versatile Ape range appeared the following year, in 1948, at the cost of 170,000 Italian lire.

The Piaggio Ape (pronounced Ah- pay and meaning Bee in Italian) truck came in a variety of guises, and if desired, lent itself to fitting any kind of custom platform, box or cube to meet specific needs.

It was designed under the direction of aeronautical engineer, D'Ascanio, who was also responsible for the Vespa scooter.

The Ape design was basically a motor scooter, with a wooden rear platform or box fitted. The driver sat atop the engine cowling. The drivetrain consisted of a differential and dual chain drive.

The Ape could achieve great things for such a small package. It was capable of hauling up to 200kg (440lb) and soon accounted for ten percent of Piaggio total sales.

In some parts of Italy, the Ape was known as the 'Lapa;' in Germany the 'Vespa Lastenroller;' Portugal as the 'Vespacar;' and in France it was called the 'Trivespa.'

The Ape could be ordered from Piaggio in a wide variety of configurations or simply as a chassis, which allowed the owner to fit whichever special body was required. It was particularly well suited for small shopkeepers and service people in crowded urban centres. Agricultural versions soon appeared for farm use. (As well as being perfect for hauling goods, the 4-passenger Calessino version introduced in 1949, soon became the ideal transport for people in sunshine-filled locations.)

The original, single-cylinder, two-stroke 98cc (6cu in) Vespa engine was soon enlarged to 125cc

**Although dwarfed by this 1958 Mack truck, the Piaggio Ape fulfilled its hauling tasks equally well, but on a much smaller scale.**

(7.6cu in) and then 150cc (9.2cu in) by 1953 for the Ape three-wheel transporter.

Like most of these micro trucks fitted with a tiny engine, a driver's licence was not necessary in many European countries. It was also easy to operate. Basically it was a Vespa scooter from the seat forward with an added rear wheel and semi-monocoque frame. The front braking, shifting and acceleration were via the handlebars. The only foot pedal was used to apply the rear brakes in case of an emergency stop.

The original Ape ultimately became known as the 'Model A' and remained in production until the end of 1954 in open scooter style with a choice of three bodies.

In 1955, a slightly improved and upgraded 'Model B' was announced and offered in four body styles.

**Left: The 4.40x8 Ape wheel and tire can almost fit inside the rim of the Mack truck's wheels with its 7.50x20.5 tires.**

**Below: On the far left of the driver, in the dash was a small interior storage bin, while on the far right was the fuel tank.**

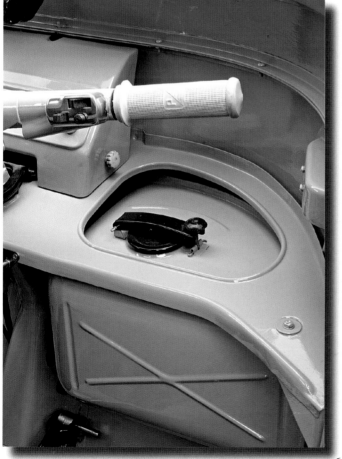

**Limited headroom necessitated assuming the traditional scooter position.**

In 1956, production of the fully open Ape ended and the 'Model C' appeared. The Ape 'C' was available for the first time in 'Cabin' and 'Half Cabin' (no doors) version. Altogether, eight different body configurations were available.

After a decade, Ape production had exceeded 200,000 units.

The stylish Model C, D and E are favourites amongst collectors.

The original design continued in production until 1971, having evolved into the longer by 76mm (3in) wheelbase 'Model D' powered by a 170cc (10.4cu in) engine (1963-64), and subsequently, the 1965 'Model E.'

**Above: The 5.8hp engine provided the Ape with a top speed of around 60km/h (40mph).**

**Left: The rear suspension consisted of torque bars and friction shock absorbers.**

In 1966 the Ape MP190 (Motor Posteriore) appeared with a direct transmission and better appointed and comfortable cabin. In 1968, the MPV became the first Ape sporting a steering wheel. It also featured double headlamps and a payload of 590kg (1300lb).

Not considered a micro truck by some is the Pentaro 250, designed to pull a two-wheeled trailer. This articulated Ape truck, with separate tractor and trailer, appeared in 1961 and was for larger, in-town uses, as well as being ideal on the farm.

**Piaggio was a successful scooter manufacturer as much for its design and styling touches, as its reliability, value and dependability**

There was no seat-back, other than a vinyl clad 130mm (5in)x780mm (30in) pad fastened halfway up the back of the cab. The pad was really superfluous, as it would require the driver to sit upright and anyone over 1778mm (5ft 10in) would have had the top of their head soon polishing the painted roof down to bare metal. Those determined not to assume the traditional lean forward scooter position, would have risked concussion from the first bump or pothole. These road hazards, would also have been impossible to avoid due to the lack of forward vision. It was a case of safety over comfort.

There were no instruments, only two warning lights on the dash and the proud wiper unit controlling the lone blade to wipe the one-piece windscreen.

**The 145.5cc (8.9cu in) single cylinder engine was placed amidships, well before Ferrari and others adopted the concept.**

### 1963 Piaggio 400

The Piaggio 400 featured has not yet been fitted with any rear bodywork. Those Apes sold as a cab only were known as a Telaio, and the all-steel cab follows the same basic design principles found in many micro trucks. On the right side of the near horizontal dash was the fuel tank, while left of the handlebars was a small storage bin.

The handlebars were placed in the centre, and thus the Ape, despite its width and comparatively generous bench seat, was a single-seater. Even the family dog – other than a Chihuahua – would be forced against the metal, unpadded cab, wall and door because the seat was a bench.

**1958 Mack towers over the Ape, which was, perhaps, one of the reasons it was not suitable for in-town or highway use in North American cities.**

A rubber floor mat allowed for easy hosing out of the interior. Your first right turn would tell you if you had overfilled the shallow, inside storage bin. When driving the noise level was excessive, but the high 'fun' factor offset this negative.

Model designations were constantly changing in the Ape model range. This version was termed the Ape 400 which indicates the carrying capacity.

Great for in-town trips to the florist, landscaper, grocers, etc for both private owners as well as businesses, the Ape was, and still is a perfect tiny hauler for covering short distances.

**Specifications**
Manufacturer: Piaggio
Model: 1963 Ape 400 (without rear bodywork)
Length: 1910mm (75in)
Wheelbase: 1730mm (68in)
Width: 1260mm (50in)
Height: 1560mm (61in)
Track: 1075mm (42in)
Weight: 227kg(500lb) (pickup)
Payload: 400kg (882lb)
Engine: Piaggio 2-stroke, 5.8hp, 145.5cc
    (8.9cu in) single-cylinder
Transmission: 4-speed (handlebar shift)
Starter/drive: Handle start
Steering: Handlebar/direct
Brakes: Rear hydraulic
Electrics: 6 volt
Body: Metal
Chassis: Steel
Suspension: Front: Helical spring & double
                action hydraulic damper
            Rear: Torque bars, friction shock
                absorbers
Wheels: 4.40 x 8in
Top speed: 60km/h (40mph)

# 6

# Autobianchi & Fiat

## Autobianchi
### Small scale style

Eduardo Bianchi built bicycles, motorcycles, cars and trucks in Italy before establishing Autobianchi SpA, in partnership with Fiat and Pirelli, in 1955. With new capital to help further rebuild after the devastation of its factories during WWII, production of an up-market model (110B), based on the Fiat Nuova 500 (110), was simultaneously developed.

The price of the Autobianchi versions were purposely set much closer to that of the Fiat 600 range, or about 20% higher to avoid direct competition for Fiat's new baby.

From 1963 onwards, Autobianchi was under Fiat control, and in 1968 was absorbed completely to become Sezione Autobianchi.

Subsequent models thereafter, were basically Fiat 500s with an Autobianchi badge. The Fiat Giardiniera had been built at Autobianchi since 1965, and along with Fiat's Furgonetta, continued in production until 1977, or two years after Fiat had ceased building the 500 sedan.

Up until 1969, Autobianchi built stylish models based on the Fiat 500.

**This 1966 version of the stylish Bianchina van was refurbished, where necessary, in Italy. It was very original except for the later outside mirrors added by a previous owner.**

Such were the good looks of the Bianchina models that Fiat considered making these cars its volume model at one point.

The Bianchina models featured more luxurious interiors, additional chrome and trim, as well as a slightly more powerful 499.5cc OHV air-cooled, two-cylinder, four-stroke engine.

Never a large volume producer, Bianchi would build just fifty cars a day at its Desio (Milan, Italy) plant, compared to the Fiat 500's 300 per day.

**Both the front and rear bumpers flow into the lights.**

**Bianchina van**

Although the 1960 Bianchina van was designed for work, it featured additional chrome, a more luxurious interior and trim, additional sound insulation and the more powerful engine.

Its appearance coincided with the introduction of the Fiat Giardiniera and the Autobianchi estate wagon known as the Panoramica (1961-1969).

The stylish Bianchina Van remained in production through to 1970. The van was promoted as "Quality transport for quality products." The brochures described it as comfortable, functional, elegant and sturdy. "The Bianchina van combines style with low maintenance cost."

It was also noted that although the Bianchina van was designed for all types of transport, " … it is particularly suitable for quick delivery service. The especially wide rear opening favours extra-easy loading and unloading of goods." Bianchi felt the van was ideal for those merchants in the radio/television, floristry, bakery, wines and perfumery business.

In 1971, a new Autobianchi Panoramica and van were introduced that were virtually identical in appearance to the earlier Fiat versions, except for badges, and remained in production until 1977.

When it appeared late in 1960, the all-steel, front-hinged, two-door design of the Bianchina van was far more contemporary in styling than the Fiat 500 version. It featured the crisper lines of the earlier Bianchina line-up, as well as the fully integrated bumper and signal light design.

There was greater overhang at the rather squared-off front end and the headlamps sat higher and protruded out over the turn signals and bumperettes. A full-length bodyline stretched from behind the headlamp to the tail lamp and vee-d just between the B-post and the rear wheel. This was complemented by a rib dividing the larger, flatter front hood in half, which also provided additional strength and stiffness.

The rear side panel was stamped and slightly raised for added strength, while also providing a perfect area for a proprietor's name.

Further added style included a distinctive faux Bianchi front grille and badges. At the rear was a lift-up, full-height, self-supporting, windowed door, rather than the side-opening door found on the Fiat Giardiniera.

Inside, the design changes were far more

A high lift-up door allowed easy loading
of cargo into a Bianchina van.

subtle with the fitting of traditional door handles and additional padding in the leather cloth bucket seats. This provided marginally more comfort and featured a different stitching design to Fiat. The steering wheel and parcel shelf differed only in detailing, whereas the dash was smooth and rounded with none of the Fiat sculpturing. In reality, the Fiat dash would have suited the sculptured exterior of Bianchina, while the curved, smooth dash would have better matched the rounded styling of the 500.

Behind the front tilting seats was a railing system. This effectively stopped any cargo from sliding forward into the front seat area. It also doubled as a tie-down, while adding slightly to the overall rigidity of the body. A similar tubular rod stretched from B-pillar to B-pillar just below the roof which also added somewhat to the strength, but conveniently doubled as a rail to hang a curtain or act as part of a framework for those wanting to install a more permanent and secure metal partition.

Behind the railing in the floor was an unlined, covered storage area of 940mm x 432mm x 152mm (37in x 17in x 6in) with a painted ribbed lid similar in appearance to the flat rear cargo area.

The load platform area of 1245mm x 1118mm x 683mm (49in x 44in x 26.9in), or 84cu m (about 30cu ft), was also painted, but fitted with aluminum rubbing strips.

**Such was the high style of the Bianchina van that, at one point, management was considering making it the volume model.**

The loading height was a convenient 508mm (20in), and numerous tie-downs were provided on the floor and sides for securing cargo.

The storage area was painted to match the exterior colour with screwed on interior side panels, painted a colour to complement the vinyl on the doors.

A second compartment at the rear, with a hinged lid, concealed the Fiat two-cylinder engine mounted horizontally.

The Bianchina van, like the Giardiniera, was powered by the now standard Fiat 499.5cc engine. The featured 1966 example was rated at 21.5hp (SAE) with an advertised top speed of 95km/h (59mph).

The suspensions on the Fiat 500 and Bianchina van differed only in stiffness, while the steering, electrical, wheels and even standard hubcaps were identical to that found on the Fiat.

**Although higher priced than the Fiat 500, the instrumentation remained very basic.**

**Specifications**
Manufacturer: Autobianchi
Model: Bianchina Van
Length: 3225mm (10ft 7in)
Wheelbase: 1930mm (6ft 4in)
Width: 1320mm (4ft 4in)
Height: 1295mm (4ft 3in)
Track: 1118mm (3ft 8in)
Load capacity: 250kg (551lb)
Storage capacity: 0.84cu m (about 30cu ft)
Engine: 21.5hp (SAE), 499.5cc, 2-cylinder,
   4-stroke
Transmission: 4-speed manual
Brakes: Four-wheel hydraulic
Body construction: Monocoque
Suspension: Four-wheel independent
Wheels: 3½in x 12in steel
Standard instrumentation: Speedometer with
   signal, oil, dynamo charge and fuel
Top speed: 95km/h

**1968 Autobianchi Furgoncino 320/2**
This rarer Autobianchi Furgoncino 320/2 was mechanically identical to the standard Bianchina van. The body appeared to be almost identical from the front bumper to the B-pillar, but in fact the factory built Furgoncino 320 had doors 889mm (35in) wide, compared to those found on the Bianchina van, which were 978mm (38½in).

The wheelbase was recorded as just 10mm (0.4in) longer with a similarly increased front track; whereas the rear track was widened to 1340mm (51in).

By moving the B-post forward and increasing the overall length to 3265mm (10ft 9in) the larger 320/2 was capable of hauling loads weighing up to 320kg (705lb) with driver.

A separate, fabricated cube 1400mmx1300mmx1010mm (55inx51inx40in), was constructed with the roof and sides made stronger by utilizing corrugated steel. Stretching (10in) above the cab, the cube was fitted with a single, left side, hinged rear opening 900mmx1000mm (35inx39in) door.

**The doors of the Furgoncino Cube van are considerably smaller than those of the Bianchina van.**

A Furgoncino 320/2 was highly functional as well as economical to operate. Whereas many micro trucks were designed specifically for city use, the 320/2 was suitable for rural deliveries and service work.

Just inside the door was the hatch panel, covering the familiar two-cylinder engine, rated at 22hp (SAE), turned on its side. Two large, round, very utilitarian-looking tail lights were fitted.

This 1968, 320/2 also featured a silver, rather than gold, plastic grille with full width horizontal spears. Inside a console was added containing an ash tray and open storage bin. There was no interior rearview mirror, as rearward vision was dependent on the outside mirrors. A full-size metal roof rack has been fitted for transporting additional, albeit, lighter additional cargo.

The Furgoncino was favoured by delivery people because of its large lockable rear space. This example purportedly saw duty delivering fruit for many years, before being sold to a construction company.

**Above: A full-size rack fitted to the roof of the cube increased cargo capacity.**

**The Fiat Topolino Type B van added functionality to the model's characteristic low running costs and reliability.**

**Although very different in appearance, in reality, much of the Type B was carried over to the Fiat TopolinoType C.**

## Fiat
### Fiat 500 Furgone and Cassone – a heftier mouse

FIAT, or Fabbrica Italiana Automobili Torino, dates back to 1899. Although Fiat began by building small cars, the company soon moved up-market to larger, more powerful vehicles.Then, in 1936, Fiat introduced the popular 590kg (1300lb) Topolino. The tiny Fiat Topolino was Italy's answer to Britain's Austin Seven. At the same time, it was not the Model T of Italy that put the Italian people on wheels. Even by the late 1950s, only one out of every fifty Italians owned an automobile.

Yet it became a beloved model and the symbol of the Italian people. Such was its popularity, the 'Little Mouse' became more affectionately known as 'Mickey Mouse.' Partly because of that, the Topolino remained relatively unchanged throughout its 19 years of production.

In 1936, the 3073mm (121in) long Topolino was powered by a 13hp, 569cc (36cu in), four-cylinder engine with the radiator mounted behind the engine. As well as being rugged, reliable and relatively inexpensive to run and purchase, the Topolino was hugely successful thanks to modern design features such as, four-speed transmission, four-wheel hydraulic brakes, and an independent front suspension.

A Topolino achieved about 160km/l at 50km/h (58mpg at 30mph). Thus, this well styled small car assured its place in automotive history.

The Furgone models were slightly modified to maintain good handling characteristics. The frame was reinforced, as well as being extended a little to compensate for the van body.

Reportedly, soon after the Topolino Type A Furgone was introduced (that designation being applied after the fact), it became apparent there were some fundamental problems. In 1938, longer rear leaf springs were fitted to further improve handling when fully loaded.

The Italian military soon put in a substantial order for the Furgone version and as a result, total Type A Furgone is estimated at approximately 12,000 units. A pickup version was also built.

In 1948, a slightly upgraded 500B was announced which included the fitting of an ohv cylinder head, telescopic shocks, and the addition of a rear swaybar to improve handling. The increase in the engine compression ratio to 6.5:1 increased power to 16.5hp. Only very minor styling changes were made to the exterior.

Furgone production has been estimated at a mere 800 units of the total Type B model output, of 21,363 vehicles.

A very different looking, slightly longer 500C appeared in 1949. Gone were the large, sloping, vertical grille, separate headlamps and full fenders, in favour of more modern, contemporary styling.

The headlamps were set into fenders that flowed into the hood which sat sbove a horizontal grille. Rear bumpers were not even offered as an option. With its basic interior and lack of aluminum trim, the Type C was a rather spartan van compared to its predecessors.

The facelift was easily and inexpensively achieved by simply reshaping the front fenders and changing the front end from the cowl, forward. A new alligator-style hood allowed for easier maintenance, but even the original front vents were retained.

Capable of 55mph and 50mpg, the last version of the Topolino continued to be popular and remained in production until 1954. It was replaced by the new, very modern 1955 Fiat 600.

### 1948 Fiat Topolino 500B Furgone & 1950 Fiat Topolino 500C Cassone

Although production of the Topolino reached 519,646 units when a final 500C Belvedere station wagon rolled off the assembly line, production of the somewhat similar vans was miniscule.

Both the 500B, and 500C had a wooden frame body, covered with shaped steel panels. Originally fitted with fixed vinyl roofs, both now have openings covered with aluminum panels, cut and nailed into place. This was done in the past to stop leaks, while also preserving the wood. The 500B still had its original, treated vinyl roof under the added panel.

Fitted on the 500C sometime in the past was a full roof rack for additional hauling.

The running boards were exactly the same on the 500B and 500C, but the front and rear fenders on the 500C models were reshaped.

Note also that both these vans had a sloped B-pillar as seen on the original Topolino, whereas the Belvedere wagons featured a narrow, vertical centre post.

Both Fiat van models were fitted with barn-style doors at the rear. Some vans were known to have been fitted with windows in the rear doors.

**The stylish lines of the Topolino were carefully maintained in the van.**

**Right:** This period roof rack was added to further increase the hauling capacity of the Topolino Type C.

**Left:** Apart from a minor difference in the fascia the dashes of the Type B and Type C were identical.

**Below:** Access to the rear cargo area was via two doors. Only the right door had a handle.

The rear styling of these vans differed only in the design of the licence plate light, tail light and door handle.

Inside, the B and C Topolino vans shared the same sliding side windows, instrumentation, steering wheel and bucket seats. Others have reported a bench seat, with the spare stuffed behind and a divider between the storage area and driver, but that was not the case in either of these very original examples. The fitting of a bench seat was abandoned in 1937, and the spare in the Type C was at the rear in a separate compartment under the floor.

The battery was stored in a compartment in the all-metal floor behind the seats.

The 500B was fitted with a wooden floor, whereas the 500C featured a metal floor with wooden strips fastened to the floor for the easy sliding-in of cargo.

*The Motor* magazine road tested a similar 500B Belvedere in January 12, 1949 issue. The British publication reported: "This type of body exacts, it is true, a penalty, in that the weight is nearly 15-per cent greater than the four-seater model, and nearly 25-per cent greater than the original two-seater, although there has been no change in track or wheelbase." The report went on to say, "… as a genuine utility car for estate

**Above: Full size, barn-style doors allowed for maximum access and loading in the tiny Topolino Furgone.**

**Below: Despite the Topolino van's functionality, Fiat designers added delicately decorative rear tail lamps.**

work, a trinity of virtues rarely to be seen, and certainly not available on any other car offering the attractions of performance, economy of running and roadworthiness which were displayed by the Fiat we had on test."

**Above left: The wooden floor in the rear of the Type B was upgraded to metal in the Type C.**

**Above right: Wooden strips on the metal floor in the back of the Type C allowed for the sliding in and out of heavier loads.**

**Below: From its B-pillar forward the Topolino van was virtually identical to the far more plentiful and popular sedan.**

Very few Topolino vans or
pickups survive today.

It has been said there are as many as ten 500B vans surviving of the perhaps fifty built, but far fewer have ever been seen.

This example formerly belonged to the president of the Topolino club and Fiat aficionado, Ezio Casagrande of Switzerland. It was formerly owned by an iron worker, who for many years used the Furgone as a daily driver. There have been pictures of three others taken over the years, but the whereabouts of these remain unknown.

As for the 500C Topolino, there is no record of how many trucks, in both van and pickup versions, were ever built, or have survived. A few exist, but some pickup trucks have been fabricated out of sedans over the decades for farm or construction use.

The van featured here is totally original. The new, brushed-on paintwork was done for photographic reasons only. This rare Topolino 500C will undergo a full rotisserie restoration over the next few years, but only after the even rarer 500B has been brought back to like-new condition.

**Specifications**
Manufacturer: Fiat, Italy
Model: Topolino 500B
Length: 3335mm (10ft 11in) Type C 3400mm (11ft 2in)
Wheelbase: 2000mm (6ft 7in)
Width: 1270mm (50in)
Height: 1377mm (4ft 6in)
Track: 1067mm (3ft 6in)
Weight: 631kg (1392lb)
Payload: 300kg plus driver
Engine: 570cc, four-cylinder
Transmission: 4-speed, manual
Steering: Worm & sector
Brakes: Four-wheel hydraulic
Construction: Body/frame
Suspension: IFS and semi-elliptic rear
Wheels: 4.25x15
Top speed: 101km/h (62mph)

# 7

# Finding & restoring

## The basics

Finding any older micro truck is no simple task. Being a work vehicle, these tiny trucks performed daily tasks until repairs were no longer viable. Many were butchered for parts to keep the survivors on the road a little longer. And, unlike cars, few were then pushed into a shed or barn with the thought of future restoration.

Once you have located a micro truck, as with any other restoration, you must find the correct shop. Specialised knowledge is required, particularly when dealing with rarer examples.

You want to pay for the quality of restoration, and not the learning curve for an inexperienced shop to locate parts, or figure out how the electrical system operates.

Yet you must also keep in mind that just because your micro truck is small does not necessarily make it less expensive to rebuild. It takes a tailor just as much time to make a small suit as it does one for a large round fellow. There may be fewer parts, but many will be scarce, and locating items can be very time consuming, which is why you want as complete a vehicle as possible to start with in the first place.

Missing bits and pieces, lenses, trim, etc, quickly add up in time and money. And keep in mind, the harder it was to find the model of micro truck you desired, the more difficult it will be to find parts or a donor vehicle. Collectors will tell you gauges, lights, and body parts are the hardest to find.

The internet and clubs are the greatest source for parts, although tracking down some fellow owners may be almost impossible depending on the micro truck.

Keen European micro enthusiasts are the greatest source for information and assistance. Dealers are also generally very helpful and professional.

Restoring a vehicle to original factory specs should not be taken too literally. JIT (Just In Time) parts is a modern production line concept. Before computerisation, when basic parts such as reflectors and mirrors, nuts and bolts, etc, ran out, others were substituted to keep the production line moving. Specs, accessories and parts were constantly changed due to different suppliers and upgrading. At the same time, when these vehicles were being built, neither the factory nor assembly line workers were thinking about collectors thirty or more years down the road trying to do a restoration.

**The cost of restoration of a micro truck can easily exceed its market value. Rarity may add value; as in the case of these two Fiats, but not always.**

A bonus to collecting micro trucks or microcars is the ability to maximise or even double one's garage space through the use of shelves or scaffolding.

In other words, not all of the same make and model of truck may be identical, but both could be correct. Also, brochures were usually printed well in advance. Artistic interpretation was often encouraged. As a result specifications, trim, accessories, and standard equipment were often different once production was underway.

Finally, it should be noted that although rare like micro cars, not all micro trucks are valuable. The cost of restoration in many cases will far exceed the market value which is ultimately determined by fellow enthusiasts. Since it comes down to the basic economic theory of supply versus demand with a bit of emotion-based desirability thrown in, buy the micro truck you want and not the one you think you can sell to make money. This is a hobby first, and a business a distant second.

## ISO

Many of the serious ISO enthusiasts doubted the existence of a Camioncino (pickup) and that the Italian maker ever made a van. Although these are extremely rare vehicles and desirable collector micro trucks, Mario Palma would be the first to tell you that ISO vehicles are not the most reliable even after a thorough restoration.

After numerous breakdowns in parades and car shows he prefers to trailer these trucks to events. His replica emergency van fitted with an ATV engine has never failed him yet.

And to quote Mario Palma, "There are absolutely no parts available for an Iso. I estimate there are only four or five Italian ISO trucks that exist and even the Spanish built examples are rare with likely only a dozen or so remaining."

## Daihatsu Trimobile

The ease of a locating a Daihatsu Trimobile is determined more by what side of the world you live on. Only a very limited number of Trimobiles were ever imported into NA and Europe.

The Asian market is the largest source, but Japan with its restrictions on older cars has resulted in most being scrapped. As a collectible in the micro truck world it has more curiosity value than monetary value.

## Moto Guzzi Ercole

As a micro truck, the Moto Guzzi Ercole was not a marketing failure, but met with limited success. Bigger than an Ape, it was not as manoeuvrable, yet could haul far larger and heavier loads. The 500cc engine was required for hauling power rather than speed. This Camioncino can be found, but also has more intrinsic value than collector value.

Not running at the time of the photo session, it took

three men to push this micro truck while my petite wife, Sandy, steered. The cost of restoration far exceeds its current market value, which makes this Moto Guzzi Ercole destined to be a lawn ornament in front of Mario Palma's museum. Only a true Moto Guzzi enthusiast would undertake the full restoration required at this point.

### Solyto
Despite the fact only 4000 units were built between 1952 and 1974, finding a New Map Solyto is still possible. There is a handful known to exist in North America, but most remain in Europe.

Frequently offered for sale through clubs and on websites, a New Map Solyto is a viable restoration project. Oozing with character, engine parts for the Ydral are available, but the Altima engine and automatic gearbox may prove to be both, difficult and time consuming to locate. The flat body panels and basic interior are easy to repair or replace.

Decidedly French in design and charm, this unique looking micro truck may not be terribly practical today, but would be a crowd pleaser at any micro club event.

### Ape
Most collectors and enthusiasts tend to associate Vespa with its built in France micro car (1957-1961) or its always popular scooters. Yet the easiest micro truck to find and restore is the Piaggio Ape with over two million having been built in the past sixty years. The fact Piaggio still makes the Ape isn't the main reason. Piaggio only supplies parts for its latest models, and no longer stocks parts for Ape models like the one featured in this book. Yet all is not lost.

As far back as 1948, the motorcycle, scooter and three-wheeler manufacturer Bajaj of India came to a sales (and then licensing) agreement with Piaggio before going independent in 1972.

The old Piaggio Ape model, based on the C, D & E continues in production and is currently known as the RE 2S. Nearly every part is available for these most-collectible versions, and is often made to the original Piaggio specifications. Also, body panels can be supplied to rebuild your Ape in either pickup or van form.

Mechanical parts are plentiful throughout the world due to the Vespa scooter roots.

Still, it should be noted that not all Piaggio Ape models from any given year are exactly the same. For example, due to German law the Ape models exported to that country were fitted with two headlamps. Other variations exist as well, usually to do with lighting and changes in specifications regarding taxation.

As the 1960s unfolded, certain safety and pollution regulations also began to be enforced which resulted in changes from one country to the next.

Still, restoring an Ape is a relatively simple and inexpensive affair with worldwide club support.

If you desire a micro truck, but dread the reliability issues often associated with an older vehicle, you can rejoice in the fact you can buy a brand new Ape or licensed-built micro truck offered in either gasoline or diesel power. (Check into the government safety and clean air regulations before placing an order.)

### Autobianchi & Fiat Topolino
Finding either of the Autobianchi trucks requires a fair bit of searching. Although the Bianchina van was a production model, few examples appear to exist today.

Despite being far rarer than the equivalent Fiat 500 van, the stylish Bianchina van is worth less. The Fiat name and 500 'cuteness' more than surpasses the Bianchi with its high style. Practical to drive, albeit a bit noisy compared to today's standards, both remain fun classics.

The lightly refurbished Bianchi van featured was discovered in Italy and shipped to Canada in 2002 along with the 1968 Autobianchi Furgoncino 320/2.

The 320/2 was a very sound example, requiring only a mild mechanical refurbishment, and it remains overall in almost original condition.

As previously noted, Fiat Topolino and Bianchina vans are almost non-existent. There are Camioncino versions that come onto the market from time to time, but be very leery. Many of these started life as a sedan, and were ultimately cut twenty years ago and fitted with a box to create an inexpensive utility vehicle. Originally, Francis Lombardi, other specialists and coachworks were responsible for a small number of very professional 500 Camioncino conversions. These would be highly prized by any collector today.

**Although the dashboard of this Type C Topolino van may not be leather covered, or require lots of fresh chrome, it will still need refinishing, re-wiring, and the instruments rebuilt.**

# Also from Veloce Publishing:

## The Lambretta Bible

**160 pages • hardback • 192 colour & mono
pictures • ISBN 978-1-84584-086-0**

**UK £29.99***
**US $59.95***

**Complete change-by-change, model-by-model, year-by-year record of every Lambretta scooter built in Italy • Full machine specifications • Frame number information • Paint code information • Many new and previously unpublished photgraphs • 1960s British dealer specials and specialy prepared machines from Lambretta Concessionaires • Original publicity and promotional material • Detailed information from ex-dealer and Lambretta Concessionaires staff**

## Scooter Lifestyle

**128 pages • paperback • 380 colour & mono
pictures • ISBN 978-1-8454-152-2**

**UK £19.99***
**US $39.95***

Scooter Lifestyle will capture the imaginations of past, present and future generations of scooter riders, and guide the reader through the scootering way of life and all its factions, giving a unique insight to the modern scene and its diversities – warts and all! Includes interviews with well known scootering personalities, over 200 colour photographs of award-winning custom scooters, best-selling scooter models, rallies and events. This book is a must-have for anyone interested in these fashionable, fun machines.

# Also from Veloce Publishing:

A full colour, A-Z reference of popular classic motor scooters and microcars from the 1950s to the 1970s, plus a hugely entertaining 'I was there' account of the original scooter and microcar era, and the lifestyle it involved.

*"It comes highly recommended, so much so that it's our book of the month."*

*Australian Classic Car magazine*

**256 pages • paperback • 441 colour & mono pictures**
**• ISBN 978-1-845840-88-4**

**UK £29.99***
**US $59.95***

# INDEX